Living In A Fractured Multiverse: The Reality Shift Effect

Roshan Cipriani

Copyright 2016 Roshan Cipriani
The spiritual and moral rights of the author have been asserted.
All Rights Reserved
No part of this book maybe used or reproduced by any means
graphic electronic or mechanical including photocopying, recording taping
or by any information storage retrieval system without the
written permission of the author.
ISNB-13:978-1523429134
ISBN-10: 1523429135

Index Page

Introduction

Chapter One

Wait... didn't he die already?

Chapter Two

Oh CERN- What Have You Done?

Chapter Three

Relocation Chess Moves

Chapter Four

In Alternate Time

Introduction

The Hadron Collider at CERN, or just known now as CERN had a catastrophic event in 2015 that may well affect the lives of humanity and the quality of survival on the earth but not a word has been spoken about in the main stream media. The media is fighting a losing battle to keep a lid on the fact that we are almost certainly in the company of beings from other dimensions, and worlds. While those running the earth have all the power, wealth, weapons and the technology to use against us, thankfully however they aren't winning. Coincidentally, if you can believe it, CERN suffered a major disaster on the same day of the Paris terrorist attacks, Friday November 13th 2015.

It is common for people with no experience or education on how real-world history and events have been orchestrated to believe that false flag attacks are not possible. They think the world's governments would have to be involved and not every nation would cooperate. The governments work together *always*, because the illusion of divisions and conflicts is just another deception on those that they control.

In order for something like that to transpire amazingly no ruse has to be too convoluted, as security forces are simply disabled by orders from higher ranking and more powerful officials.

As in the past the media was ready with the story and prepared to broadcast and go into print the minute the happening was reported as completed.

Fears that high energy collision experiments with the LHC, the world's largest particle accelerator positioned in the Jura mountains neighboring the Franco-Swiss border, could activate an Earth-wide or even universe-wide catastrophe have been anticipated and of course would need to be contained by the media in order to limit or avoid panic thereby continuing to maintain control.

Although we are told that the data collected from the collision procedures would help scientists to check their ideas about the structure of the universe, they do not really have any idea what could and most likely will occur.

While I do believe from all my documented research that CERN is being used for malevolent purposes, G-d the Creator is still in control of all things and will continue to be.

Walter L Wagner and Luis Sancho brought a lawsuit on March 21, 2008 in Federal District Court in Honolulu, Hawaii, seeking a temporary restraining order to stop CERN from proceeding with the LHC. They had requested that an assessment of risks be conducted. They believed that the LHC could generate a magnetic monopole. The physicist Paul Dirac suggested that particles could exist that hold a single magnetic charge instead of two.

What do they know that we don't?

Even though I doubt this accelerator can come close to producing some of those unknown forces that could destroy our creation; fears have been expressed that if monopoles exist and are generated by the LHC, the consequences could be cataclysmic.

To cause even further questions of their intent there is a giant statue of the Hindu deity Lord Shiva outside of CERN's headquarters dancing the "cosmic dance of death." Is this purely imagery, a coincidence or something else entirely?

This seems not accidental when you learn that some in the scientific community have stressed the likelihood of opening "parallel universes and extra dimensions."

Awakened humans, who do not join the evil ones in their desire for destruction, demonstrate the failure of their plot for absolute domination. Consciousness is a unified effort among all beings on earth which does clearly include the prayers of the faithful who want good for all.

Like most stories in the Bible, the Tower of Babel episode is often misunderstood and is taught as an alteration. The tower was an attempt not to rise to heaven but to take down any connection between heaven/G-d and man.

Once again we have another group here on the earth trying to stop our natural reconnection to the highest dimensions, and just like in the past they will ultimately be stopped.

Today we are facing possible rips in the time space continuum, blown open gates between the spiritual realms and the physical dimensions, and structural fractures in the fragile areas between the dimensions; or the space between the "tick and the tock."

While at first it may seem fantastic, the scientists at CERN have openly admitted they are attempting to open other dimensions and they do not know what will be brought into or taken away from this one.

These are only some of the concerns that we should have with the simple fact that scientists just don't know what they will find should the experiments have even limited success.

CERN is definitely more than a particle collider given just the electrical force they are employing. I do believe that the black hole risk at LHC is real and physicists are playing Russian roulette with unknown forces and the people of the earth. Many people have expressed the belief that if their reservations turn out to be correct, the entire Earth and every person here may be gone in seconds.

Yes, gone but to exactly where? Then again what if the damage already done has been more subtle, but just as catastrophic?

Since everything in the universe is, at its core pure energy as in electrons, protons and neutrons it stands to reason that if what I've learned about these massive machines magnetic fields could turn out to be a lot more destructive than what is going on in the experiment itself.

Most people have no idea how much energy there is in the Universe. Or how much is in the entire Cosmos. Are they so senseless to really believe that we could generate enough energy to somehow destroy the Universe?

I don't believe that we can destroy it, but I do believe it is possible to significantly alter somethings and create things that we will not be able to fix or handle.

CERN is experimenting with uncontainable powers of physics in which they acknowledge they might fracture time and space. Then to make it more interesting Stephen Hawking has put forward that such experimentations could go perilously out of control. The "atom smasher" is now functioning at its maximum level in an attempt to discover miniature black holes, which are considered an important sign of a "multiverse." Since June 2015, the energy that the LHC uses to bang particles together is twice what it was during the time when it made the detection of the Higgs boson. While they believe that parallel universes can exist within these dimensions, they feel that only gravity can leave our universe and go into those extra dimensions.

However at least two of the stated major objectives for CERN is to collapse and breakdown the "God Particle" that generates and sustains our physical world and to rip a hole through the covering that is the obstacle protecting our physical reality from the unidentified, non-physical dimensions thought to be found external to our physical universe.

So if CERN does eventually destroy matter, and everything in our universe is matter, what does this actually mean? Extinguishing physical matter eradicates the limitations and barricades produced by physical matter that preserves us from inflowing the non-physical and other physical universes into our own. We then would be open to just about anything from anywhere.

But what if this is exactly what they have already done with dimensional reality?

The Mandela Effect is a theory of parallel universes, constructed within the idea that many people have similar alternative memories about past events, and they may all have been in a different dimension with different occurrences and outcomes and not be mistaken in their recollections. This is because the dimensions may have fractured at some point in the past, allowing people from one to occupy another and that therefore not only do parallel, inhabitable universes/dimensions exist, but we have somehow been moved around like pieces on a chess board.

The Mandela Effect was first designated as such online in 2008/2010. A woman named Fiona Broome discovered that others had a distinct and different memory from the norm, and it was similar to hers. She along with many other people remembered that Nelson Mandela had died during his imprisonment in the 1980s and was confused as to how he could have now died again in 2013. This revelation opened the conversation and people began revealing other strange recollections not always shared by people around them, yet something they have in common with many other people in the world.

What could account for this? Is there any explanation for the occurrences of certain things, events, places and people that numerous people remember in a different way and precisely?

For me as well as many other people the multiverse dimensional theory is used as a concept to account for it.

Perhaps people randomly moved to nearly identical universes, taking the place of their alternate self without ever realizing exactly what has happened. They may feel strange and some may even notice things that are not quite right, but without ever realizing that it could be something else just dismisses the occurrence or feeling as insignificant although puzzling.

My belief is that the universe /dimension you are in is the one where your primary or main conscious mind and physical body is placed. So any incident where you narrowly escaped death for example, happened that way because you continued to want to live and was somehow placed into another dimension believing you are still alive in that same time stream. However in that other alternate universe you did actually die at one of the terminating points or traumatic events.

Now consider the altering of this entire thing by the scientific community and their experiments going back to the 1940's, and its effects on the nature of reality.

Yet for most people the easier answer is the stress-free one. Simply accept the idea that your memory and those of many people is extremely faulty sometimes on the exact same things. You just dismiss this all as nonsense and ignore anything else that seems out of place to you.

This is for many people a lot easier than the inkling that your entire life and everyone you know is just an extremely precise copy and you are somehow now living in an alternate dimension. A dimension where things you recollect are now vastly different, and the "mistakes" in your memory even though they are detailed and precise and shared by millions of other people should just be ignored.

I had no idea I was not the only one. I thought this was just happening to me until I discovered that there are other people experiencing this as well.

Less important memories — whether someone went out for dinner on Saturday or Sunday, where they parked the car, or how much change is in the glove compartment — are more easy to write off as nothing much when they're not accurate, precise and detailed, and when they are not a part of your personal narrative and life story.

Now after years of researching this subject, it's exceptional for anything to take me completely by surprise, but it gets more bizarre every day.

Greater, more worldwide recollections — the ones listed in this book— are diverse, and this is a problem. For me I am now aware that my memories don't match my current reality dimension or time stream.

Most people are okay with these things, whether they understand them as a quantum consequence, or prefer to think they misunderstood or misremembered in the first place.

However I have learned to not simply discount anything, particularly with something as bizarre, fascinating, and occasionally even alarming as the Mandela Effect.

Chapter One

Wait... didn't he die already?

When lines from movies, book titles, television show names and even people's names are not what we previously have known them to be, we could be seeing evidence of this multiverse right in front of us. When landmarks that never existed to you previously are now said to exist and are hundreds of years old, and when whole cities are completely different than you know it to have been it becomes very hard to just ignore it.

The way I see it, we seem to be able to shift in between different dimensions/realities and they don't seem like they are correct or "normal" somehow to our own experiences.

Although we have a very powerful and recognizably magnificent personal instrument; our Brain, which we still do not completely understand, we are willing to assume the "mistake" must be with us, and our thinking on some massive scale.

The world is a big, mysterious place, and it'd be foolish to believe we have all the answers, and now only these two choices. Our understanding is limited, but many of us who have experienced these things know something is radical altered and different.

Could the time streams of different parallel universe/dimensions have been altered by the scientific experiments done previously? Could scientist somehow have fractured a time stream and now affect several dimensions by the circumstances occurring now with CERN?

As we all know there is NO WAY to prove the time stream has been altered, just as there seems to be no possibility of bringing or sending anything physical out of one dimension to another as evidence to prove a point.

Perhaps memories of that other time stream occasionally surfacing from time to time in our brains and multiple people experiencing this at the same time is the evidence of the change and the alteration.

For the past 5 or so years it seemed I was drifting between realities/parallel universes, without understanding what was happening. The only thing I knew for certain was that I didn't have that bad of a memory of my own existence. My recollections were precise and heavily detailed, and I was in severe shock on seeing things that I had already lived perceived differently. Yes, sometimes they were small things and it seemed unimportant, even minor but those little things still mattered to me and to those who remember differently.

There are many of us who are convinced that what we remember is real and we witnessed things that now we are told just never happened.

I definitely believe I was in a parallel universe and I detected a difference almost immediately, but I didn't understand what happened. The one great thing about this is that you can't use doubt to disprove the truth of someone else's memories.

It seems that the people from a different dimension/reality/time stream for some reason brought their memories intact with them. If you remember everything exactly as it is now you are in the dimension/reality in which you have always existed.

My intention is not to change your thinking with this book, but to alert you to what has recently occurred among more and more people worldwide.

You might be surprised as to how effects like hyperspace, physical portals, and multi-dimensional existences can be affecting our daily lives. Yet this phenomenon has all become quite extensive since the CERN project in particular has been in operation. Once something is changed in the past, it changes the future. Entire things will be completely different; not partially different.

When two dimensions collapse, or collide the dominant dimension seems to take over. Perhaps time streams were not just separated by time, but by particles of matter and these are the differences in physics itself, and somehow CERN has interfered with that.

Technically speaking, there would have been an infinite number of progressive dimensions, each individually divided by little alterations. But due to the fracturing of the time streams; events and things have changed and people are in strange and unfamiliar dimensions with strangers that look like the people they knew, and situations that are vastly different than their previous reality and recollections.

What if CERN has inadvertently left remains of prior histories on current time streams, or has split entire time streams in half somehow?

Then there is ITER. This is the International Thermonuclear Experimental Reactor and is an enormous fusion reactor being constructed by 35 countries in southern France. ITER is also constructing a neutral beam test or NBFT site in Padova Italy. Strangely very few people will have heard of ITER or what they plan to do. While CERN actually began operations in 1954, ITER is still a decade away.

In 2008 ITER and CERN signed a Cooperation Agreement to collaborate not only in the arenas of technology such as superconductors, electromagnets, cryogenics, control and data procurement and composite civil engineering, but also in organizational areas such as funding, purchasing and human resources.

This collaboration includes software programs and working closely with DEISA, in full Distributed European Infrastructure for Supercomputing Applications; a European consortium of national supercomputing.

DEISA also maintains a network link with an agency known as Tera Grid – another supercomputing network in the US.

The ITER project has been branded as the world's paramount human attempt and illustration of world collaboration since the incident of the Biblical tower of Babel. This is so much more than it would at first appear; ITER has even created its own multi-national currency called the IUA. Why would a "science project" need its own currency?

The simple statement that a concept like the Large Hadron Collider with its stated aims even exists should be a warning signal, that evil is behind it. Human minds basically will not create a device such as this for no reason other than to "see if something happens."

If they are willing to spend billions of dollars on these huge contraptions, it must be something of an unimaginable magnitude that they are attempting to do.

I believe there are many parallel universes and perhaps they could in theory collide with each other and become somehow fractured, resulting in the transference of some type of magnetic energy and with it some of the people from one dimension/ reality into another without anyone realizing at first what might have occurred.

There are many people who feel that CERN is modeled after the pineal gland or third eye because it's a vehicle for a variety of time travel mechanics. There is of course another theory on why this is happening, and it is that it was intentional.

Maybe someone time traveled to aid in changing the future. They made slight changes because large changes would be too apparent and whoever controls or has authority over this sphere; be it some regime or clandestine social group would realize this had happened and could have possibly rectified it before any changes became observed. I have noticed that almost all the things that are changed are very public things that would go unobserved with the exception of ordinary regular people who have read, saw or utilized these things on a daily basis, and so would definitely have detailed recollections that would be hard to just ignore.

You cannot speak for anyone but yourself, so at the end of the day both recollections of the same event would be accurate, since it would be a personal experience. For me this is not philosophy, or speculation this is physics.

I know that the word "multiverse" is used instead of "universe" because scientists have accepted through quantum mechanics and the string theory that there truly are more than one dimension and more than one universe.

These are some things I have noticed in my original time stream of this multiverse that are completely different in this relocation one. See how many of them resonate with you as well.

Muhammad Ali is still alive (early 2015) but I recall his 2009 death, and the funeral on tv.

Frank Gifford died in 2012. His 2015 death was actually quite a surprise.

I Remember Betty White's death in 2014, It was announced that she died–"the last of the Golden Girls is Gone." Imagine my surprise that she is alive and well in 2016.

I remember Nelson Mandela having died in South Africa in prison and His wife Winnie Mandela later became the first black female president of South Africa.

I remember Scotland as being separate from England - its own island and was in the North Sea.

I remember having conversations with a family member that other family members say doesn't exist. Funny thing is he was at my wedding and brought a great wedding present.

Some people myself included remember Ronald Reagan dying in 1999, when he died again in 2004 in this 2015 time stream.

I learned in school that 2 bombs were dropped on Japan: Hiroshima, and Nagasaki however in other timelines it was 3 bombs.

I remember the death of Whitey Bulger in 2013, and a documentary on his life, yet in this time stream, he's still alive.

Mongolia is a gigantic country on global maps and a major world player in many people's reality. For me Mongolia was a province in China. Now in this reality time stream/dimension it is a large country between Russia and China.

I recall the company name as always "Proctor and Gamble" not "*Procter* and Gamble."

The comet that was the talk of our lifetime was called Hailey's Comet- Here it is Halley's Comet.

The Forrest Gump movie I saw when was first released is totally different than what we have here.

The hit song "Straight Up" by Paula Abdul sounds completely different today than I remember it when it was released.

There was actress Doris Day's dying in the late 2000's yet she is still alive now.

Gone with the Wind's Scarlett O'Hara's famous line: "Wherever Shall I go; whatever shall I do," is here stated differently.

Both my daughter and I remember Cheesecake Factory restaurant meals jokes about the food being horrible and the joke was you just ordered the food to stave off the sugar shock from the magnificent desserts. However, in this time stream it is considered a great place for a good meal and people rave about the food.

I remember being taught about the 52 states. Children born after 1990 seem to all remember 50 states.

I recall John Goodman's death from a heart attack shortly after the Flintstones movie in 1994, but in this time stream he has lost weight and is alive in 2015.

I remember Korea being SOUTH of China near Vietnam, certainly not out North next to Eastern Russia.

I remember that the Lindbergh baby had never been found, yet here it is reported as having been found dead.

The movie line in the Wizard Of Oz was "Toto, I don't think we're in Kansas anymore." In this time stream it is "Toto, I have a feeling we're not in Kansas anymore."

Reba **McIntyre** is spelled **McENTIRE** now- which is different from the Scottish ancestry of McIntyre in my remembered time stream.

I remember Vladivostok being much more North in Russia and not bordering Korea as it is shown on maps in this time stream.

I remember watching on television that event in Tiananmen Square in which the Chinese young man refused to get out of the way of the army tank and was run over and killed. It was shocking and everyone was speaking about it. In this time stream that never happened.

Living In A Fractured Multiverse: The Reality Shift Effect

Some remember **Fruit Loops** breakfast cereal in the 1960s, yet I only remember it first appearing in stores in the late 1970s. And it was always spelled *Fruit Loops*, not **FROOT LOOPS** as in this time stream.

I remember the air and fabric freshener product *Febreeze*, in this dimension it is *Febreze*.

In my original time stream/ dimension there was no animal called a Narwhal. This is apparently a whale with a long horn on its head like the unicorn. In this time stream it is called "the unicorn of the sea." Growing up I watched nature programs on TV–Jacque Cousteau , Marlin Perkins, David Attenborough, and others and never heard of this. I thought it was some kind of a joke since I've never heard of nor seen a picture of a narwhal before 2016.

I recall Easter Island as having been discovered by James Cook/ Easter Island, and I remember him finding it uninhabited. Rapa Nui is the name of what I knew as Easter Island, given to it by its native people, who have continually inhabited the island for nearly 3,000 years in this time stream/dimension.

I remember the sun being bright yellow, not white, and I learned in science classes at school that there were only 4 or 5 cloud formation types , but in this time stream clouds appear in odd shapes and forms and there are over 20 types here.

Thanksgiving was always on the third Thursday of November in the United States. And in this time stream, it's the fourth Thursday in November. It stands out in my mind because my grandmother taught it to me as a child, when I learned the countdown to Christmas day.

I remember the peace sign become popular in the 1970s; it had the arms facing upward; never downwards as in this time stream.

I remember the GREAT Pyramid of Giza being off into the desert MILES away not literally 700 FEET from the suburbs of the city of Cairo Egypt, as it is here.

I remember that Jane Goodall died and was remembered for her research on gorillas, when in this time stream she is still alive and famous for her research with chimpanzees. Gorillas in the Mist was a movie which had a TV premier and I distinctly remember the movie I saw was about Jane Goodall; staring Susan Sarandon. In this time stream Sigourney Weaver is the actress in that movie and it is about Diann Fossey.

I remember the pictures of this massive white statue called **Christ, the Redeemer** overlooking the city of Rio de Janeiro on a gigantic white rectangular base. Now it is just a large statue. The base has also radically and mysteriously changed to a smaller base and is a black square cube.

I remember a BBC America Television show called MI-5, however in this time stream it is called SPOOKS, and while still about MI-5, it was never called that especially in the US.

Cartoons were **Looney Toons** now **Looney Tunes** and **Merrie Melodies** in this time stream, yet I knew it as **Merry Melodies** my entire childhood.

I remember a peanut butter known as "Jiffy" the original brand name. So when I saw "Jiff" peanut butter I thought it was a name change by the company. It seems that at least in this time stream there was never a name change and it has always been known as "Jif." However I remember my brother and I being very insistent with my mom when we were children that she only buy "Jiffy" and not "Skippy" another brand of peanut butter. I even remember the song from the commercial.

I remember *Oscar Meyer* as a deli product company, in this time stream it is *Oscar MAYER*. I even remember singing the song in the commercial...about "my bologna has a first name it's ---- O-S-C-A-R, my bologna has a second name it's----- M-E-Y-E-R....!"

I also recall that all traffic lights were green yellow and then red at the bottom, so I was surprised when I noticed it in reverse.

I also remember the spelling of words being completely different. I spelled a word as "suprise" now it's "surprise" and "lightening" instead of "lightning", and "realise" is now "realize." I was always big on reading and writing and had entered spelling contests every year as a child. I paid attention to words and I am a writer now, so I find this bizarre. We were taught the proper grammatical usage is "my brother and I," now here in this time stream it is "my brother and me."

Here combined words are non-existent: 'infact' is now "in fact"; afterall to "after all"; overall to "over all" moreso to "more so;" alot to "a lot"; alright to all right; and no-one is "no one."

"Dilemma" is remembered as being spelled "dilemna" and "dammit", as "damnit" The spelling of the nation of Columbia changed to Colombia.

The colors chartreuse and puce have switched here. I remember Chartreuse a pink -reddish purple, not puce's yellowish-green color.

The automobile symbols are different also. Volkswagen – VW, here has a space between the two monogram letters, and Volvo in this time stream has an arrow added to the circle, making it the symbol for "male,"and not the circle missing a piece that I remember.

Vancouver Island seems larger here and British Columbia is much larger also on these maps.

I remember New Zealand being one land mass. In this time stream it is now two islands and it is bigger than Italy.

The Bahamas were NEVER just off the coast of Florida in my time stream, only Bermuda was. Cuba was NEVER that close to Mexico.

When I visited NYC years ago Manhattan Island jutted out into the Atlantic. The statue of Liberty was on an island a little farther out into the Atlantic and not near New Jersey. You had to take a ferry to get to Staten Island as they never had a bridge.

Here in this time stream I learned there are 4 bridges to Staten Island. I had no idea that there was any bridge. I always thought that you had to use a ferry to go to Staten Island.

I do remember in the movie *Working Girl,* actress Melanie Griffith had to ride the ferry back to Staten Island and I clearly recall the scene. In this time stream the movie does not have that scene.

Martha's Vineyard was a district on Long Island. It has been moved away, leaving the Bay in Long Island, here Martha's Vineyard is an island.

I recall Sri Lanka being directly South of India, not off to the East of it. ☐ I was shocked to see Gibraltar moved from the strait between Spain and Morocco –to be on the East coast of Spain.

I was stunned in particularly by South America's 1000 mile eastern shift, out of what I recall as the straight alignment with North America.

The JC Penny Store in this time stream is JC Penney.

American Television chef of "Bizarre Foods" was Andrew Zimmerman, here he is Andrew Zimmern.

The host of the Twilight Zone television series was known as Rod Sterling, here he is Rod Serling.

Walmart was ALWAYS a blue logo- never Wal*mart in white logo in my original dimension.

The Talladega Superspeedway and the Daytona Raceway were in Florida, however in this dimension the Talladega Superspeedway is in ALABAMA. I was shocked that there isn't even a town called Talladega in Florida.

Then again there is that Rock of Gibraltar. It is British owned. It is in my time stream/Dimension, a source of contention between England and Spain because Spain believes due to its proximity to their coast it should be considered as Spanish territory even though it is **an island offshore.** This is how I remember it; an island of disputed territory, not a part of the land mass of the country of Spain, sitting surrounded by water facing Morocco.

I have the same memories and many more as do some of my friends and family; it seems that almost everything is incorrect, countries and landscapes, words, people in the public eye and often historical events from what I have always known. ☐

If you don't recall things in this way, *you are from THIS dimension/reality/ time stream,* or, alternately from *yet ANOTHER alternate dimension/reality/time stream* from my original one.

My belief is that these are dimensional shifts or fractures that we were experiencing are variations in reality which is different than our memories.

Some horrible technological tampering has done something peculiar and it is affecting everyone, some I believe just haven't noticed yet, or are putting it down to being mistaken and shrugging it off.

It may be that two or more dimensions have collided and somehow like glass tubes they are fractured no longer sealing in one reality from another right next to it.

It's scientifically INEVITABLE that when a large group of people remember something they knew for certain, it's true for them. And I mean by a large group of people – MILLIONS OF PEOPLE, who have experienced the scenario. I believe that there must be something more here than meets the eye.

When people say the difference in what I remembered is just a memory lapse I realize their dismissive attitude is meaningless when MILLIONS of other people also recall it differently and with the same miniscule details that I do.

I am still amazed over the geographical changes on the world map. It bothers me immensely as I vividly remember a land mass being called the North Pole it was never a large lake. New Zealand was above Australia and it was one land mass and not in two pieces.

Australia is now half the size I remember and is missing part of its shape at the top. Indonesia and Australia are much closer to each other in this time stream.

Australia in my original dimension/time stream reality was out in the ocean completely isolated and very far from any other land mass.

It happened to me again, my reality shifted I believe around 2013, in the midst of continuous disturbing personal activities.

However even before that time, I remember in 2008 waking up one morning and I got out of bed and went to my daughter's room and stared at her while she slept to make sure "she was still there" and I went back to bed. I woke up some time later, seemingly okay and I wondered "now why did I do that?" I did not understand but I felt that something was very wrong and I was uneasy for many days after that.

I do remember the sun use to be brighter before that day and the sky used to be very blue and crystal clear not opaque and dull.

Perhaps I was moved somehow, to stop the events that were happening; because of the stress ordeal and other related issues in my life. But the how and why I have no idea. I am still asking myself why is it so different *this time*?

My own personal time stream has changed drastically since 2001. We were always close and seemingly overnight there were deep issues with my immediate family, and complete isolation from one another. I remember being very scared all the time then because I felt that "this was not my life." When my reality somehow changed my memories seemed different to what I was seeing after 2001.

Everything seemed different and even friends seemed not the same people I knew before. I continued to say out loud to no one in particular that "sometimes I just don't resonate with *these humans*." When I'm feeling as though I'm not like the people around me, I can also feel the strong difference when I met a *"regular"* person like myself. There may be some validity to something having happened when you hear of people who seemed "out of character" and then they disappeared out of daily life without a trace.

I also have distinct memories of three near death incidences where I either just woke up, found myself outside on the ground, or realized I was sitting somewhere and did not know what happened as I did not recall how I got there but I thought "at least I'm ok now."

Maybe I've died at least three or more times already, and with each death I wake up in a different dimension/reality.

The idea that Einstein is actually right; and that when you match the frequency of the reality/dimension you want you go into, then that reality/dimension becomes your new reality/dimension is really quite astounding when it may be happening to you.

Now I know for absolute certain that something is seriously going on. I feel the necessity to "detail check" with my daughter and close friends to see if they continue to share my reality dimension/time stream memories. At first I tried to ask subtle questions without drawing too much attention to why I was inquiring about the simplest things. Now for some time I have also had the thought that my reality may be unlike someone else's reality.

Many people have had strange, paranormal experiences with a piece of clothing missing, or being a different size or color or just popping up in strange places. You simply cannot just write all these realities off as the result of massive bad memory, or brain lapses due to age or stress.

So many people can recall funerals of famous people only to learn they are alive and didn't die. I feel it's because some of us have moved into a different universe/dimension and time stream where these people haven't died yet.

That then explains why some of us remember deaths while others do not. We have no explanations, but there are many suspicions.

The cause of all of this could be the hadron collider at CERN and its work on dimensional portals. The scientists have ignored the fact that when something changes, it changes not just here or there, but EVERYWHERE. You enter a completely new time stream of consciousness where the whole past is different. Consequently when you go back into the past, it's has completely changed.

That's why there are no newspaper clippings, paintings, advertising commercials, and even your own DVD movie collection that you can access to verify your own memories. It has all changed when you did and the possibility exists that you may have done more than physically move.

 Any two realities that share a timeline close enough can stream-slip together. Anything that is the same between the two will merge. There is so far no way to say how that is determined. It's apparent though, because if someone was reminisced as dead but realized to be alive, undoubtedly the reality dimension/time stream on which the individual is still alive has taken dominance.

Thinking through even my own rationalizations for what I have experienced, I feel sometimes as though someone has tried to trick or program me, or as though I'm in some type of sick mental experiment. To me reality is so *unreal* and as I get older this feeling intensifies.

So for anyone who's new to this idea — and slightly astounded, identifying many things as different now – yes, it's okay to feel unstable.

For me the question isn't whether this happens, but how it's happening and why.

Chapter Two

Oh CERN- what have you done?

We like to make-believe we know what we are doing with science, and that we have it all thought out. Seeking in the darkness will always have the possibility of being a drawer full of sharp knives and no matter how carefully you shove your hand in you will get cut sooner or later.

I still believe, in some way, the colliders from FERMILAB on up to ITER are involved, possibly along with a share of the ever present digital technology, which could have speed everything up more than the analog technology of the past.

There is a well-known disaster theory which pictures a precarious universe, and the Mandela Effect may represent even bigger dangers, a perilous multiverse of dimensions and time lines no longer isolated from each other.

Then there is AI; artificial intelligence which is the development of computers that are "awake and conscious" and superhumanly smart. I believe this could most likely be a scenario with large computer networks like DEISA which also maintains a network link with an agency known as Tera Grid in the United States.

When greater-than-human intelligence motivates developments, that advancement will be much more rapid.

If the Technological Singularity can happen, it will not even be acknowledged by its creators for what it is. The fact is, whether their intentions are malevolent or benign, humanity basically does not have enough evidence to correctly calculate the consequences of their experiments.

However when humans depend on calculations from technological machines for making choices and decisions, these same technological machines, are able to determine outcomes, factor in probability for their favored outcomes and basically control humanity.

Is the obvious explanation each time true? No, but this doesn't mean the obvious explanation is never true.

Then there is invisible mind control.

Scientists across the planet, not just the United States, are fearful of the work they are doing in relation to mind control technology; and, many of them believe that his technology is ready to be released. In fact, there have already been beta tests done.

These are just a few of the patents that have been granted and the inventions that may be involved.

US PATENT 6,017,302 – SUBLIMINAL ACOUSTIC MANIPULATION OF NERVOUS SYSTEMS.

In human subjects, sensory resonances can be excited by subliminal atmospheric acoustic pulses that are tuned to the resonance frequency.

The 1/2 Hz sensory resonance affects the autonomic nervous system and may cause relaxation, drowsiness, or sexual excitement, depending on the precise acoustic frequency near 1/2 Hz used. The effects of the 2.5 Hz resonance include slowing of certain cortical progressions, sleepiness, and disorientation.

US PATENT 5,935,054 – MAGNETIC EXCITATION OF SENSORY RESONANCES

USED TO CREATE EXTREME AGITATION OF A TARGET, WHILE TARGET IS IN HYPNOTIC STATE.

US PATENT 4,858,612 — HEARING DEVICE MICROWAVE HEARING

The multiple frequency microwaves are then applied to the brain in the region of the auditory cortex.

US PATENT 6,011,991 – COMMUNICATION SYSTEM AND METHOD INCLUDING BRAIN WAVE ANALYSIS AND/OR USE OF BRAIN ACTIVITY

USED TO CREATE SADNESS, EMOTIONAL CONFUSION, FEAR, PANIC, ANGER.

US PATENT 4,877,027 – HEARING SYSTEM

This employs microwaves in the range of 100 megahertz to 10,000 megahertz that are modulated with a particular waveform that consists of frequency bursts.

Each burst is made up of ten to twenty uniformly spaced pulses grouped tightly together. The bursts create the sensation of hearing in the person whose head is irradiated.

US PATENT 5,270,800 –SUBLIMINAL MESSAGE GENERATOR

The user can control the length and spacing of the subliminal messages to create different sentient inceptions.

US PATENT -6,506,148 – NERVOUS SYSTEM MANIPULATION BY ELECTROMAGNETIC FIELDS.

This technology may be utilized by a video stream, and can be transmitted as an RF signal.

This is just a sampling of what may be bombarding our minds and senses. CERN's LHC activity as well as all other collider activities across the earth coupled with HAARP/Tesla/Geo -engineering technologies uses electromagnetic energy that can have devastating effects on humans as well as on the atmosphere of the earth. Could these frequencies be a type of frequency trigger?

We are literally submerged in an ocean of electro-magnetic radiation. Is it possible that we have somehow inadvertently technologically interfered with the dimensional frequencies and time streams of other planes of existence?

Could every dimension/time stream consist of its own distinct frequency?

The possibility does exist that when we began to cross the threshold using massive quantities of subliminal atmospheric pulses of energy into our space, we altered other distinct frequencies.

The more technological changes such as cell phones and WIFI have changed our frequency and our perceptions, the more this would align another dimension/time stream and their frequency to somehow merge into ours by matching that frequency.

Then there is the theory in which many people feel that they are not in their respective original "home" dimension /time stream, after they had a life-threatening occurrence where they might have actually died.

Within this theory, it is further suggested that when the disaster was inescapable, people were somehow taken from that original "home" reality/dimension time stream and placed in the adjacent "comparable home" reality dimension time stream. Often it is the one they find themselves living in quite uncomfortably, and constantly questioning.

For me personally the permanency of some of these memories proposes that I am from or have lived in the same or very parallel dimensions/realities time streams, before finding myself here. I can feel something is wrong yet I can't put my finger on actually what it is.

If we take one interpretation of quantum studies accurately, *every* possibility creates a new universe dimension/reality.

So, in at least one reality, people are fully aware of their having a life threatening problem and being "removed" to another dimension in which the situation was better.

It would also mean that there is at least one other new universe dimension/reality in which people are being "removed" to another dimension to escape a possible inevitable death and they are totally unaware of it and just continue living as usual.

They may just remember everything from the old original "home" reality dimension, and only question themselves when finding large groups of others experiencing similar recollections in perfect alignment with their own.

If you haven't experienced the Mandela effect or don't react to it as I have, this may not make much sense to you. While many of the changes are very subtle, it has had a surprisingly emotional effect on me.

 Well, I thought I was losing my mind, so I began writing down all the things I began to notice that were completely different than I clearly remembered. I tried to find a pattern or patterns but it simply was not there for me.

I had trouble grasping the changes, although I believe that those who have strong memories of an occurrence seem to preserve those memories, while others embrace the memories of the new reality/dimension time stream.

We have noticed cell phone messages were either changed, never received, or content differed showing altered wording for the both the sender and receiver. Often we would receive a text message hours after it had been sent.

I have heard things clearly while lying down and about to sleep to the point where I have been in conversations that I am unable to recall on awakening. No, this was not just me dreaming as I could definitely feel the difference. Moreover I have had more than three NDE experiences, and seem to have experienced more precognitive dreams than seems normal.

For me I learned the idea was first put forward by a graduate student at Princeton- Hugh Everett in 1957. In the "Many Worlds" explanation, instead of collapsing from superposition into a single reality, the wave function branches into multiple realities comprising each possible outcome. This understanding of quantum mechanics conveys with it the fantastic allegation that all possible pasts exist, each contained within its very own universe. So that every time a decision is made, another complete universe branches off from this one.

The inference is that, among the endlessness of universes being created, we will all follow the precise branch that ensures our immortality. That doesn't mean you never died. In that other universe, your family mourned your demise while in this one you lived as though you just had a close call.

Supposing this quantum immortality is valid; there could be multiple physical universes like our own existing side-by-side where *consciousness does not end at death.* The idea of quantum immortality violates no known laws of physics and quite scientifically is supported by those same laws.

The first law of Thermodynamics specifies that energy such as the electrical charges produced by your brain, or the heat your body puts out - cannot be created or destroyed, but only alters in form—implying that the energy that powers your body must go someplace when you die, and since our consciousness cannot be destroyed, but is infinite you have to continue to "exist."

I remembered a car accident I was in when I was in my early twenties. I remember that the car was completely demolished from being rear ended by an eighteen wheeler/semi-truck. My car was crushed all the way up to the driver's seat from behind. People at the scene couldn't believe I was able to get out and walk away from it with just a sprained wrist and a headache.

I have thought for many years that perhaps I actually was killed that day and that I somehow ended up being given another chance or a do over. EVERYTHING in my life changed after that event. I even left the state I lived in and relocated to an island. To this day, I am not certain I survived in the reality/dimension I had previously inhabited, although at that time I just felt I had received somehow a second chance. Then I had this happen to me again, I feel as though I died in 2013 and was somehow given another opportunity.

Throughout my life I have had many such experiences where various scenarios occurred and where I felt after it "almost happened" that thought of "hey, that was close."

Some say that irrespective of the cause of death, if the many-worlds explanation is true, then there will *always* be at least one branch/dimension or reality where the "miraculous survival" circumstances has occurred, and that version of "you" will never die. Quantum immortality could mean that you can never start down a road where you end up in a situation with a no chance of survival. This would also mean that by being close to one another, two people would be less likely to die in the other's world by excluding any mishap except one that could only take you both together.

It seems that there are forces in the Universe that are regulating these events.

At least some physicists believe that the Large Hadron Collider - CERN the most powerful particle accelerator every built, is being prohibited from attaining full power by the universe itself. Malfunctions, including power problems kept delaying projects continuously. Perhaps this is the universes way of granting us a chance at survival in this dimension.

Many of these events occurred in the electric arc that produces the beam of protons. Because the events became more frequent as the intensity of the proton beam increased, they "are expected to be very critical for LHC operation at higher energies."

The Adiabatic Quantum Computer is in fact artificially intelligent and this computer is connected to the LHC at CERN. The scientists at CERN with the use of this computer are planning to carry out a project termed "AWAKE." It is publicly slated for the end of 2016; however I believe that the activities at CERN have been directly causing the shifts in what we call the Mandela Effect.

Consciousness, the state of being self-aware or awake is something physicists have not come into agreement on concerning a descriptive theory but have continued to delve into. Quantum Brain Dynamics (QBD) is the theory that states there is a new quantum field that is accountable for consciousness. If this is true, it would signify that our personal consciousness survives by means of communication with a quantum field, or due to contact by quantum particles.

If the scientists at CERN discover a way to explain, influence, and even create consciousness on a quantum level, it seems to me that no matter what, there will be consequences to such an action. They may intend one thing but by the nature of the technology they will get other things as well. Just as the nature of the spinal cord joins the brain with the rest of the body, and continuously lights up with messages like a fiber optic system it too is capable of much more.

It's an expressway of physical information and also the super highway for spiritual information. A light carrying particle is also a dark matter particle. The scientists at CERN are now actually watching for the Particle or the atom of "Evil," in the same way they sought for the "God" Particle.

Without the ability to keep humans in an unconscious state, humanity will be waking up. One would think that if this super collider is successful in altering or creating consciousness, even more people will remain asleep and in a unconscious or controlled state.

There is however another idea regarding all of this and it is so much more diabolical.

This is perhaps the time to use our DIS-**_CERN_**-MENT? Perhaps just like the spinal cord and the collider there is a double message here. It is a double meaning as we have are now feeling very CON-**_CERN_**-ED.

This project and its divisiveness has become a pivotal idea in humanity's awareness and consciousness. This is even more sinister when you consider that quantum mechanics says that our focus of awareness and consciousness actually is what creates our reality. If even one human mind can create reality from pure belief, then it is possible that someone will attempt to gather a large group together to harness that power and direct it in areas of their con*cern*.

They have constantly stated that the activities at CERN are to eventually amalgamate religion, folklore, and science together as one. For the modern world folklore would be our television and movie culture of storytellers.

Perhaps it is our consciousness that they are seeking to control and harness and our separate dimensions time streams/ reality that they are "colliding" inadvertently while doing this. I believe that they are attempting to change the external world by harnessing and using our collective consciousness. But what if they had this plan all along?

So I wish to offer this idea.

In one of my favorite movies – "The Party" (1968) with Peter Sellers, where he is playing an INDIAN actor in America in a film on British India; there is a guest list to *the party* that he accidentally gets placed on.

Rather than run out of town and blacklisted for destroying the movie set he was supposed to be working on, he is invited to the studio head's home for the dinner party and one of the guest on the list is *Mr. Bernard Stein.*

Bernard is an alternate anglicized version to the Germanic *Beren* similar to the *Beren*/stein- *Beren*/stain bears of the Mandela effect memory duality.

There is also another scene, in which Peter Sellers is the actor that has the director cut a scene because he is wearing an "*underwater watch*" which was totally *out of the time period that the film was being made in.* In the later scenes there is a conversation at the party in which a guest tells a story about *his watch being stolen.* Then there is the actress/ love interest of Peter Sellers character. An up and coming singer who is to *star* in a movie for the studio - Michelle Monet who sings in the party these lyrics: "Both you and I have seen *what time can do.*"

At that juncture there's the culminating scene of the movie. A baby elephant at the party led into a pool full of soap bubbles, and young people, that take over the entire house. The elephant has words painted on it that says "The World Is Flat."

While the *Russian* dancers are involved with the out of control entertainment, the entire house is practically destroyed *with water*, as the *Indian* actor and the *French* singer leave together early in the morning.

It is also very interesting that as the guest enter the house where the dinner party is being held, *they must walk on a very narrow entry foyer constructed over running water.*

However the main character Peter Sellers clumsily steps *down into the water*, even losing one of his shoes in the process.

Could this be basically a flat earth and moon deception disclosure by subliminal programing, indicating stolen time, time travel, being out of place for the time period, a memory marker such as a children's book, the *republican* powers for *the elephant* being washed up of everything by the pool, and the statement that the "Earth is Flat," and *surrounded by water* finally being admitted by the powers that be.

Or perhaps I'm just reading too much into an old movie.

There are some who believe that we are time travelers, visiting this time stream/reality from the future, who have forgotten so they must try to leave us clues in the contemporary culture; such as a movie.

There are many ideas and theories, but one thing is true; if we exist in one reality and we witness changes to this one reality, then something is causing the changes either intentionally or accidentally.

What if these perceived redone/alterations are made to our memories when we sleep? If you're not sleeping when everyone else is, or if your sleep cycles are totally different, is your vulnerability to the "alteration of memories" also different, resulting in your memories not corresponding with the new time stream/ reality majority memories?

I am fully aware of how the mind can be tricked into hearing, seeing, or remembering things that didn't exist, but all of this information together has brought back very vivid detailed memories of more changes than just these small differences in memory.

I decidedly believe CERN and the activities of those scientists to also be involved in the activities that have brought about the things that we are calling the Mandela Effect for want of a better term.

Most people have no idea of the true threat of CERN and how it could be affecting their daily lives so that the idea that something they are using constantly is now also a product from CERN will usually come as a great surprise.

I was quite surprised to learn that the World Wide Web (www) was designed *in this dimension/time stream* in 1989 by Tim Berners-Lee; a British scientist at CERN. Here the *Web* was originally conceived as a way to enable spontaneous information-sharing amongst scientists in universities and organizations across the world. It was publicly made available for communal society usage on April 30, 1993. This is totally different to what I knew about the creation of the internet in my remembered time stream/reality.

However here there is approximately 96 percent of the *Web* that is restricted and inaccessible to the general public. This is known as the "dark web" and in order to gain access to it specific programs that allow entry must be accessed that are restricted from the general public. This never existed in my original reality and I find it a strange and seemingly unquestioned circumstance.

Makes you wonder what exactly goes on there and is it impacting people using computers daily without them being aware of its use?

Then there is that humming sound so many of us are hearing. It sounds like a very distant idling heavy truck, or a generator just vibrating with the motor running. I found I can't block it out even when using earphones; therefore it appears to be picked up somehow by our bodies.

I didn't notice it much during the day, and there were times when it was barely noticeable or even not there. When my air conditioner was turned on I couldn't hear it, as the air conditioner unit was loud itself. However as soon as the unit turned off, the humming sound seemed to have resumed.

I did notice that during the times I personally heard and felt this I had intermittent nose bleeds; although I had never connected the events.

Even though I moved 2500 miles away from the first location where I was constantly hearing the "humming vibrating sounds;" it has not made any change in my experience. In researching this I found many others are able to hear this also. It seems that so many people of all ages can be affected by this phenomenon and yet we have no explanation. Although the reported occurrence among children seems to be very low at present.

Could it be related to the activities of LHC at CERN? Perhaps when CEFN was activated it resulted in a form of destruction of something within our universe as projected by Stephen Hawking. Maybe there are people who were "moved" by divine involvement possibly to a parallel universe. Then there may be a merging of parallel universes into one another taking specific people from different dimensions/realities into this different dimension/reality or even into several others.

The CERN LHC was first used on September 10, 2008. After this transpired, earthquakes occurred in Iran, Indonesia and Japan of over a 6.0 magnitude, and the next day the system shut down. Perhaps the shock wave from the energy beam dumping disseminated through underground bed rock devastating already hot quake zones, and those plates slipped creating a cluster of earthquakes on that day.

On November 20, 2009, the CERN LHC was employed again and the Norway spiral anomaly was seen in the skies a few days later on December 9, 2009.

Some people speculate that this spiral in the night sky may have been a wormhole opened by the combined result of the CERN LHC working in connection with HAARP.

For those who don't know; Project HAARP is High-frequency Active Aural Research which was based in Alaska at that time. HAARP is one of more than three comparable facilities in the world. One of them is in Norway and is known as (Eiscat) and another is in Russia and is called (Sura). The HAARP project turns the ionosphere into a laboratory.

Their project was to heat the ionosphere with antennas, in turn generating ion clouds that will duplicate an optical lens, which will mirror ELF waves to the earth. This throws an unprecedented amount of energy into the ionosphere, focused to a point in the ionosphere.

According to the patent - the work of *Nikola Tesla* in the early 1900's was made the foundation of this research.

On April 25th, 2015 scientists at CERN were starting the arrangements to power up the LHC machine. The power volumes were brought up four isolated times in the moments before the earthquake on April 25, 2015 in Nepal.

If the CERN machine put energy into the circling particles to the equivalent of 15 kilo tons of TNT or more, before dumping them, and did this four different times in a row then an earthquake was not only possible but probable.

While correlation does not equal causation, Nepal is on a major geological fault in the Himalayas where two tectonic plates connect and was most likely the most vulnerable to that concentration of energy. The LHC at CERN presents the biggest magnetic field on Earth, next to the one centered in the earth itself. So it is theoretically an earthquake machine that disturbs the magnetic field and dumps energy into the ground.

It is interesting that Nepal is the historically termed "the land of Shiva," and the statue that is installed outside the building at CERN in Geneva is of Shiva. There are spiritual connections involved in this work, and implications that we haven't even thought of yet.

Throughout my entire life I have examined everything. I became a writer in part, due to my thirst for examination and not taking things at face value; but searching out and weighing the facts.

It is a fact that when CERN put the LHC on within a week the (8.8) Chilean earthquake occurred, which was the sixth biggest earthquake in history. Then after 3 months CERN was operational again and we experienced the fifth biggest earthquake in history. It would be improbable that with all of the evidences pointing to CERN's involvement, we could just overlook two of the biggest earthquakes in history in a year of CERN's operation.

Then there is the Norway Spiral anomaly or portal. This was probably created by the EISCAT ionospheric heater facility in Ramfjordmoen, Tromso, Norway. It is the scientific mother of the United States military HAARP program. EISCAT is an abbreviation for the European Incoherent Scatter Scientific Association.

It operates three incoherent scatter radar systems, at 224 MHz, 931 MHz in Northern Scandinavia and one at 500 MHz on Svalbard. They are supposed to be used to study the interaction between the Sun and the Earth. Thousands of people watched as this spiral appeared in the night sky, grew larger and then developed a large black hole in the center that apparently swallowed the entire anomaly.

Since the technology at HAARP can create spirals in the sky and there is an ionospheric heater at the Ramfjordmoen facility near Tromso, Norway the realization that this antennae array was exactly where the blue beam pointed down left no doubt as to who or what may be at work here.

In 2006 there was also a spiral anomaly wormhole in Tomsk Russia, one in China and later on one in California. There has also been one seen recently in Queensland, Australia.

There is also a HAARP facility located near the small town of Vasilsursk; the Sura Ionospheric Heating Facility.

Some say its activities at that time may have brought about or intensified the fury of Hurricane Katrina that devastated New Orleans in the United States. Whatever the cause of something on the material plane, it does not negate or even define its significance on other levels of consciousness.

The reason the earth has a magnetosphere is because the earth itself has a geomagnetic field. The Magnetosphere around the earth acts as a protective shield.

According to CERN's official website, "All the magnets on the LHC are electromagnets. The main dipoles generate powerful 8.4 tesla magnetic fields – more than 100,000 times more powerful than the Earth's magnetic field."

These men want to go where no one has ventured before and are supposed to be very shrewd; it now seems however that they lack the wisdom of the Creator and are finding this out.

Their madness is destroying the workings of earth's magnetosphere and leaving us to be blasted by gamma-rays and neutrinos from local supernovas, pulsars and neutron stars which could smother our atmosphere and biological ecosystems.

While the fluctuating actions of the sun, including sunspots, solar flares and coronal mass ejections, are acknowledged to cause alterations in the magnetosphere, it seems clear that the undertakings of people shouldn't – but what if they are?

And what if their real aim is exactly what we have been experiencing?

It seems that any two realities/dimensions that share a time stream close enough can merge together. Anything that is the same between the two will amalgamate. These would be the closest in proximity and presumably having diverged from each other previously in their time stream dimensional differences.

Therefore when the magnetic anomalies alter these separate frequencies in the dimensions by being more than "100,000 times more powerful than the Earth's magnetic field," it could be exactly what they were aiming to produce.

Chapter Three

Relocation As Chess Moves

To believe, or not to believe. - Hamlet

Humans can be deceived and lied to by other humans causing them to do and think irrational things based on what they have confidence in because of their experiences. It is appalling how the powers that be have lied, damaged past history and then revised what they desired to spoon feed multitudes. The curtain of lies however is slowing being removed. I believe that these things are attempts to stop our discovery and awakening.

I've come to comprehend that humans must be more spiritually powerful than we have been led to accept as true. And the NWO select few's plan is to also make sure people are unable to improve and recover their true selves by dumbing us down through GMO's, fluoride, injections, manufactured sicknesses, chemtrails, amalgam poisoned dental fillings, and the other evils of treatments in western medicine.

You need to begin with a clean slate and start thinking for yourself; because most people have been trained not educated. Ask yourself – "are you really ready?" Only a small amount of research is required. Identify and remove all the deceptions and deceivers, and then just keep going.

At the end what will be left is the TRUTH you have been looking to find.

The so called scientific community is today run by deceptive and dangerously mad men.

They built the LHC at CERN. They have scientist that are working for them whether they understand that or not, doing "experiments" in which they fire subatomic particles at speeds approaching the speed of light and have them smash into each other.

This according to their own literature will generate what is identified as antimatter, which in layman's terms is the reverse of matter; which comprises our entire universe. Antimatter shouldn't occur in these realities/ dimensions, because when it comes into interaction with matter, it's highly volatile.

It has the reactive destruction equivalency to destroying all existence. Antimatter is not only explosive, but it is associated with mega earthquakes, cracks and fractures in our reality/dimension, and is believed to contribute to dangerous violent behavior and psychosis in people who are in close vicinity to it.

While the magnetic shield covering the earth is there for a reason, re-creating any part of what does not belong here on the earth can have devastating effects.

There has been no proper environmental impact assessment studies or unbiased safety calculations of the LHC experiments.

So while the LHC and the scientist are using our home of over six billion people as free lab rats while ignoring that there are risks of dangerous consequences we are experiencing anomalies in our daily lives that seem to get stranger on a daily basis.

I believe that the Creator G-d can stop the activities at CERN anytime He chooses to do so. I also believe in Universal law which requires certain things to just continue to play out. Perhaps these actions by these evil minded men are the equivalent of chess moves they are attempting to play with the Creator and the Universe itself.

I believe the evidence is there for the existence of multiple dimensions and time-streams, multiple harmonic universes that are analogous yet reliant on each other. A reality/time stream dimension shift of some kind is and has occurred. Millions of people can't just have delusions on a complete line of precisely the very same mistaken memories.☐

Most people don't want to believe they could be vulnerable or even be programmed in such a way as to have a hard time recalling events and things to their minds. However, that is exactly the way mind control programming works.

While there are many theories on what is happening, any evidence of it being the way you remember will never be found, because everything changed at once, except for the memory. This could not be attributed to mind control programming.

While the theories are many, I'm 99.9% certain the parallel realities/dimensions theory is correct, and we are now living in a fractured multiverse.

We know that all possible arrangements of particles exist at all times, and we see results from that arrangement, however something is strange. I've had items that I know be present disappear from right next to me, and when you don't know what's happening. It's really frightening.

But I totally do remember Madagascar being different. Madagascar was further south, approximately east of South Africa. I remember it being mostly uninhabited and a protected nature preserve. So imagine my surprise to learn that the population of Madagascar was estimated at just over 22 million people. I remember it in my original reality/dimension and it was also much smaller. It had no human inhabitants, towns or governments, only scientist and researchers. It also had no flag.

Then there is Cuba. It is at least ten times as large as I remember it being - 75 miles to 760 miles - and much closer to the Yucatan Peninsula than ever seemed possible.

It dominates the Caribbean here and, having grown up in the Caribbean and flown all over it visiting and living on many of the different islands for over 30 years, it is quite astonishing to me.

All past globes and maps now reflect our current time stream/dimension and seem to be the only "evidence" available. On the other hand, islands and continents being in different locations represent a profoundly different Earth, which would've needed to start being altered plainly billions of years ago.

It is the massive geological differences that I clearly recollect from our current world, that I think about now. My understanding of alternate dimensions/realities is that the more recent the time stream branching, the more identical the worlds/reality/dimensions should be. This makes sense to me as they would have more in common. The farther back we branch off, the more opportunity for difference.

So this occurred billions of years ago, when the earth was formed, in order to have geological variations in different realities in a physical dimension.

Now it has taken a long time for us to have noticed that our dimensions have divided into multiverses. It still doesn't explain what we are experiencing here and at present.

A time or reality shift has occurred but the evidence of such things as I understand them is that we'll see much of the world the same as before. However, we should expect to see changes in history, people, events and geography.

I've always felt that humans are spiritual beings having a physical experience; the Mandela Effect seems to endorse it to me. I can't deny my memories and make-believe that this is the reality/world dimension I have been living in, this is not the reality I remember. It's a feeling that is very similar, if not exactly the same, as the one you get when you're having a déja vù experience. Suddenly everything is artificial of what it really should be. Nevertheless this feels really bizarre. I want to somehow wake up. It seems we are long past the point of denying that some of us are experiencing altered realities.

It's particularly remarkable to me that my daughter and I share a number of alternate memories, and yet also have a number of dissimilar recollections the other doesn't share. We are both surprised that we both ended up here; aware of this occurrence, and accepting of it because we are appreciating the positive things that happen here. I have no doubt that parallel dimension exist, whether or not they explain the Mandela Effect.

One thing I have noticed is the rate of "changes" has been increasing to where it appears to be almost a daily occurrence.

Maybe, this is just the start of a succession of events that will transpire, which in consequence will completely change reality.

Maybe in order to deal with the changes mentally, we each need to a matter of becoming a "better" person... in action and thought. I think often about what I left in my original time stream/reality and dimension. Who did I leave behind and how are they now?

Some of those people here just aren't the same. Have they noticed that I'm either not there, or different? Did I die and so not expected to be there now?

Here they familiar often seems like they are completely different people, just similar in appearance. Is that what is happening in my original time stream/dimension with a different me?

When everything is unfamiliar, perhaps the change is only with you; the observer? Or are we living in a "people moving universe?"

What about dreaming? Is it possible that our dream state can cause our consciousness to slip between realities/dimensions in a similar way as quantum immortality suggests? Have you ever dreamed you had an accident and woke up injured? It has happened to quite a few people with seemingly no rational explanation?

Everything has data programmed in it – that information can be revised with advanced technology. Awareness of that technology has been around for a very long time, until now it was principally veiled from the masses for thousands of years.

If there is an underlying self-correcting encryption to the universe, then one might identify its existence in mathematical descriptions of field interfaces. Our observation that alternate presents exist is exactly in keeping with what is widely held by physicists when they say we exist in a superposition of states. Its mechanisms can be understood by researching legends and myths, mathematics and some of the earliest structures built on the earth like the pyramids. It seems there is meaning in these things–that hidden symmetries exist and continue, and there is a greater design in all that is.

For me I'm determined to see that the new reality dimension time stream synchronicities act correspondingly in manifesting positive outcomes for ordinary as well as significant occasions in my life. It's sort of a - "when life gives you lemons, make lemonade" - idea. I don't know anything for certain, but I believe these ideas are interesting.

Chapter Four

In Alternate Time

Back in 2013, the television show "Rewind" was slated to air, however due to reasons not released, the show never aired past the first episode notwithstanding the fact that the pilot seemed to be an enormous success.

The episode that aired of "Rewind" showed what happened when a disaster devastated a major city, and a group of people, including one civilian scientist and two military specialists, go back in time to stop an incident that could stop the event. There's a concern about probabilities and butterfly effects, but they seem to be able to get past that and still use untested equipment that looks shockingly like a replica of the LHC at CERN all the while stating they don't really know what will happen when they do.

The dialogue for the show even has a character state, "we are standing in the core of the most powerful particle collider ever built," and their intention is to travel back in time to alter past events in order to change the future and avoid a terrorist attack.

I have to admit that the unnerving visual similarities between the LHC at CERN and the "time travel - particle collider" in this movie were shocking.

Speculation is that it may have been cancelled because there is a similar government operation secretly sending people through time, and they are using "the particle collider technology, "to do so. It leaves one to wonder if the show was cancelled because it unknowingly revealed too much information and placed another use for the LHC; that of time travel, into the public perception. The fact that human perception is powerful is a well-hidden fact from the public.

Since total collective consciousness affects everything; time, reality and space, and they do not want any "thought interference" in what they have been doing from the masses.

Since we create the future with our thoughts, and since most people do not think, but only react to the outside world; they are unaware of exactly how powerful our minds really are.

Since the phenomenon of the Mandela Effect touches the perception at perhaps the quantum level, science has neither the equipment nor procedures needed to measure impartial evidences. I believe that there are some occurrences in life that cannot be definitely explained by science/scientific assessment. However that does not mean it is not actual and significant. One reality may have been pushed into another timeline affecting the future and changing some of the recent past.

I'm torn between two distinct intellectual certainties; that this can't be real, and that it must basically be my faulty memory and a clearly existing statistical confidence that it is all happening and very real.

How we react and perceive time, space and the very nature of existence can be what can animate the actual multiverse and fuel changes.

Yes, something is tainted in our present world and more and more people are sensing it all the time. Some term it a spiritual emerging, but I wonder if it is an awakening to the awareness that our very being has been transformed and we are existing where we were never before. There is the possibility that we are people both metaphorically and factually "in alternate time." Alternate here means to switch back and forth, and I believe that is exactly what has been happening for many people.

Something has transpired, and we have been moved into a different time stream/world/dimension. I believe time travel is possible, but only by entering other instances of the multiverse where time remains fixed.

I could have crossed back and forth many times in the few seconds it took me to write this paragraph but any differences would be so minor or affecting that I probably just haven't noticed.

My theory is that the more different two dimensions are, the farther apart they are in the multiverse and the more energy it takes to transport from one to another. Consequently the less likely that someone will naturally/inadvertently shift from one to another without some outside assistance.

The undertaking of affecting the movement of tectonic plates alone never mind the ages of time involved impede any possibility of humans nor any earthly powers being the agent for changes to what I've known my entire life. So many things have changed from what I've known them to be my entire life. Some are routine like the spelling of certain words being completely different, to the locations and topography of New Zealand, Cuba, Japan, Australia and all of South America for example being drastically different. I'm intellectual enough to comprehend the allegations of the impossible acts that would have to have been undertaken to transform the world as I've recognized my entire life to the world that is before my eyes now.

Outside smells completely different. Everything tastes different. Driving through the small town I'm now living in, I look and it seems that the colors are completely off. Green isn't as green as I recalled it to be.

I notice it mostly in blues and reds; they seem less warm and more flat. Things seem to lack "harmonic energy" and I haven't seen "live" colors recently.

Things seem a little less solid than I remember them. The grass seems an element less dimensional to me; sort of weak. I felt somewhat flowing and detached from the world around me.

If this phenomenon lasts or increases a lot more people are going to find themselves in a world they no longer are familiar with. For instance, what if you wake up one day and the town you lived in no longer exists but you have clear memories of being there your whole life. What if your spouse or parents don't even remember there was such a place?

The possible proof of alternate parallel universes has been now presented by an astrophysicist at Caltech. Scientist Ranga-Ram Chary found what he called a mysterious glow, looking back in time to just after the Big Bang more than 13 billion years ago. He studied the light left over from the early universe, the so-called "cosmic microwave background," and said the glow could be due to matter from a neighboring universe "leaking" into ours." He further stated that, "Many other regions beyond our observable universe would exist with each such region governed by a different set of physical parameters than the ones we have measured for our universe."

It appears that what we think of as the universe is a part of an infinite whole that has no beginning and no ending, therefore looking back in time, could only occur by some form of *time manipulation* because what you are seeing actually happened BEFORE you saw it except when view the night sky of course.

The idea is that any time you look up at the night sky and see stars you are looking back in time, seeing the light from stars as they were thousands or millions of years ago. However, when there is evidence of enormous amount of radiation apparently from an unknown source, it may be an indication for the existence of parallel universes if the information is studied and confirmed.

In Physics the String Theory envisages parallel universes, yet the problem has always been discernable proofs. Quantum physics is still a mystery. So either idea has yet to be validated.

So another idea that can be put forth is that if there is instead only one earth one dimension/ time stream reality; perhaps they inadvertently opened a black hole with the LHC at CERN and it took the earth through it somehow. Instead of destruction they completely changed time and reality.

This is why so many people who were dead are alive again, and the geophysical changes that are a fixture have been reformed. This would be a new reality that was created artificially.

This is just another possible theory as an explanation for what might have occurred.

It seems we will never really know a because any evidence of things being the way you remember them will never be found in a new reality where they cannot exist. Either everything changed at once, except for the memories within our minds or we have been moved somewhere else. Could it be that we were moved to a place where it really doesn't matter to anyone what we know and believe in the face of what we are seeing and being told is real and has "always" been?

Personally, I have wondered about certain turning points in life, such as making one decision against another and now consider perhaps if this is when we created these other realities. Could it be as simple as when we make choices, everything is then birthed? Could each additional reality split off of one reality every time there is more than a single possible choice or outcome?

Could these realities or dimensions be damaged somehow just shattering and in a fractured state drop into the nearest similar reality, taking on the features of the enduring reality?

Could this be why some individuals do not just integrate appropriately with a new reality, and have fragments of the fractured dimension/ reality to remain in their memories?

Think about the consequences of that. We live in a quantum entangled world. It all could have just changed because it just changed. It doesn't need a cause. It is different because the universe decided it would be different. This has enormous insinuations for reality as a whole. How do you even begin to comprehend how immensely joined everything is with this? Our dimension/reality or time streams can continue to change in an instant and then back again.

There is a higher power in charge of all of this and that explains how tremendously multifaceted our existence really is. There's someone observing all of this, and making decisions on how things occur. The good news is that that higher power, the universe, also known as God is in control.

Other Books By Roshan Cipriani

Rise - Be True To Yourself-Inspire Others To Live
How To Get Through Any Wall In Your Life
Train Up A Child – A Scriptural Guide To Parenting
The Art Of War For Parenting Your Teenage Child- How To Win A War You Didn't Even Know You Were In
The Key To This Life - Conscious Faith In An Unconscious World
Destiny – Past Present Future
The Seven Pillars Of Wisdom –A Sabbath Celebration Guide
Life Lessons Learned
In The Fire – Accessing Miracle Power During A Crisis
The Kingdom Lifestyle - Living By Faith And Not By Sight
God's Secret Wisdom –Principles And Secrets Of The Kingdom Of God
The Greatest Principle - The Kingdom Of God And Biblical Economics
Bricks Without Straw- Spoiling Egypt And Spoiling Babylon; The Mighty Wealth Transfer
When Failure is Not An Option
Real Faith – How To Have It And Why It Matters
The Bibles Healing Promises
I Say What They Said- Miracle Bible Prayers
The Psychology Of Stress-Dismantling The Enemy's Weapon Now
Never Quit-The Secret To Getting Through Any Wall In Your Life
His Poetry Store
SMALL BUSINESS SUCCESS- How To Write A Book Every Weekend
The Seventy Two Lunar Sabbaths- Sabbath Observance By The Phases Of The Moon
BUSINESS PLAN: Make God Your Partner –He Commanded His Blessings
PROSPERITY CONSCIOUSNESS – Living In An Abundant Universe (Personal Biblical Economics) Volume 1
Metamorphosis-Mirrors Of The Soul, Awakening To The Real You
Waiting In Goshen
How To Be Smart And Have Common Sense
None Of These Diseases –Sickness And Genocide In Second Egypt
Patience To Inherit The Promises- How To Stand By Faith Until Manifestation
The Lord Is My Shepherd, I Shall Not Want- Personal Biblical Economics
DIVORCE RECOVERY: How To Live Again
UFO COVER-UP: Biblical Evidences Uncovered-(Conspiracy) Volume 1
12 Easy Vegetarian Recipes-Healthy And Inexpensive
TRAVEL: How To Behave On An Airplane
NINJA SMOOTHIES: 21 Green Weight Loss Smoothies For The Ninja Professional Blender
DALS: 7 Simple And Healthy Recipes For Indian Style Beans
RICE: 7 Cheap And Easy Recipes For Indian Style Rice
CHUTNEY: 9 Simple And Healthy Chutney Recipes
CHAI: 7 Quick And Tasty Chai Recipes
Second Exodus From Second Egypt
Asset Protection And Wealth Management-Volume 1 -Trust And LLC For Legal Asset Protection
RELATIONSHIP RESCUE FROM THE BIBLE: What The Bible Says About Relationships

www.ingramcontent.com/pod-product-compliance
Lightning Source LLC
Chambersburg PA
CBHW081207180526
45170CB00006B/2244